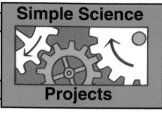

Simple Science Projects

PROJECTS WITH

WATER

By
John Williams

Illustrated by
Malcolm S. Walker

Gareth Stevens Children's Books
MILWAUKEE

For a free color catalog describing Gareth Stevens' list of high-quality books, call 1-800-341-3569 (USA) or 1-800-461-9120 (Canada).

Titles in the Simple Science Projects series:

Simple Science Projects with Air
Simple Science Projects with Color and Light
Simple Science Projects with Electricity
Simple Science Projects with Flight
Simple Science Projects with Machines
Simple Science Projects with Time
Simple Science Projects with Water
Simple Science Projects with Wheels

Library of Congress Cataloging-in-Publication Data

Williams, John.
    Projects with water / John Williams : Illustrated by Malcolm S. Walker.
        p.   cm. -- (Simple science projects)
    Rev. ed. of: Water. 1990
    Includes bibliographical references and index.
    Summary: Provides instructions for projects involving rafts, paddle boats, and other objects that float.
    ISBN 0-8368-0771-5
    1. Floating bodies--Experiments--Juvenile literature. 2. Hydrostatics--Experiments--Juvenile literature. 2. Boatbuilding--Juvenile literature. [1. Floating bodies--Experiments. 2. Experiments. 3. Boatbuilding.] I. Walker, Malcolm S., ill. II. Williams, John. Water. III. Title. IV. Series: Williams, John. Simple science projects.
QC147.5.W56 1992
532'.076--dc20   91-50549

North American edition first published in MDCCCCLXXXXII by

Gareth Stevens Publishing
1555 North RiverCenter Drive, Suite 201
Milwaukee, Wisconsin 53212, USA

U.S. edition © MDCCCCLXXXXII by Gareth Stevens, Inc. First published as *Starting Technology — Water* in the United Kingdom, © MDCCCCLXXXXI by Wayland (Publishers) Limited. Additional end matter © MDCCCCLXXXXII by Gareth Stevens, Inc.

Editor (U.K.): Anna Girling
Editor (U.S.): Eileen Foran
Editorial assistant (U.S.): John D. Rateliff
Designer: Kudos Design Services
Cover design: Sharone Burris

Printed in the United States of America

2 3 4 5 6 7 8 9 9 97 96 95 94

CONTENTS

Words printed in **boldface** type appear in the glossary on pages 30-31.

FLOATING

Some things float and some things sink. It does not matter how large or small they are. Big, heavy things can float, while small things may sink.

Very large boats like this one take big, heavy loads all over the world.

Testing for floaters

You will need:

A small tub filled with water
A piece of cardboard
Felt-tip pens

1. Collect objects made from different kinds of materials. Rubber, wood, metal, and plastic are good materials to use.

2. Draw two circles on the cardboard. Make the circles overlap. Label one circle "Floaters," and the other "Sinkers."

3. Test your objects in the tub of water. Do they float or sink? Write the name of each object in the Floaters or Sinkers circles.

4. Test your floaters again. Do they always float or can you make them sink? Push them under the water. What happens? Do they fill with water and then sink? In the overlapping part of your two circles, list any of the floaters that you can make sink.

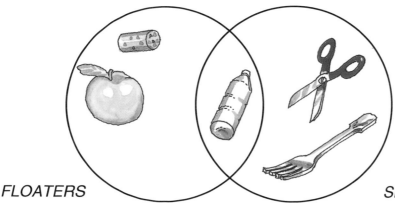

FLOATERS   SINKERS

Making aluminum foil boats

You will need:

Aluminum foil
Scissors
Modeling clay
A bowl of water
Weighing scales

1. Cut out a few sheets of aluminum foil and make some simple boats. Experiment with several shapes. The boats should all be small enough to move around in the bowl. Make sure they do not leak.

2. Put the boats in the water and put small pieces of modeling clay into them until the boats sink.

3. Weigh the amounts of modeling clay in each boat. Does each boat hold the same amount of clay?

4. Now weigh an empty boat. How much lighter is it than the clay that was in it?

Making a clay boat

You will need:

Modeling clay
A rolling pin
A bowl of water
Marbles

1. Use the same amount of modeling clay that you used for sinking one of your aluminum foil boats. Roll out the clay until it is thin. Make it into a simple boat.

2. Gently lower your boat into the water. When it is floating, carefully fill it with marbles, one at a time. Count how many marbles the boat will hold before it sinks.

SEE IF IT FLOATS

Fill a sink with water. Try to hold a block of wood under the water. Can you feel the wood pushing upward? It is this upward push that helps objects float.

These big logs float. These men are able to stand on them in the water.

Experimenting with floating

You will need:

A bucket of water
A brick
String
A **spring balance**
Modeling clay

1. Tie the string to the brick and hang it from the spring balance. Gently lower the brick into the water. See what happens to the reading on the balance.

2. Do the same experiment with a ball of modeling clay. Like the brick, it seems to weigh less when it is in the water.

3. Roll out the same amount of modeling clay and shape it into a boat. Hang the boat from the spring balance and lower it into the water. What happens to the reading on the balance this time?

RAFTS

Long ago, people used rafts to travel along the water. Rafts are floating platforms made out of logs or planks secured together.

People also hollowed out tree trunks to make canoes. These canoes rolled over easily, so people put **outriggers** on the sides. Even today, outriggers are used on some boats to prevent them from tipping over in the water.

*These modern canoes, or **kayaks**, are used for sports.*

Making rafts

1. Make a simple raft by tying two corks together with rubber bands.

2. To make a larger raft, put some Popsicle sticks across the corks to give them support.

3. Put your raft in water and see how it floats.

Making an outrigger

1. Make a model person with pipe cleaners. Sit it on your twig boat in some water. See if you can stop it from rolling over.

2. Use three wooden sticks to make an outrigger for your twig boat. Arrange the sticks and the twig into a square shape with a rubber band at each corner.

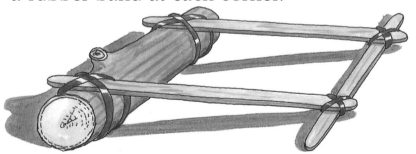

3. Put your model person on your boat now. Does it roll over?

LONGBOATS

Longboats were used hundreds of years ago by the **Vikings**, who were famous for their sea journeys. They often sailed hundreds of miles in very bad weather. They even sailed across the Atlantic Ocean, from Europe to North America.

This is a model of a Viking longboat. It has a small sail and many oars. Count how many oars you can see.

Making a model longboat

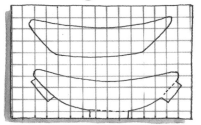

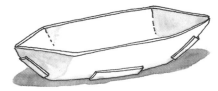

You will need:

Cardboard
Pencils
Scissors
Glue
A wooden stick
Paintbox

1. Cut the two sides of the **hull** from a large piece of cardboard. On one of the sides, include three small flaps. You might want to draw these shapes on graph paper first.

2. Attach the two sides of the hull together and glue the flaps to secure the hull.

3. Cut out the other parts of the boat from the cardboard. Cut out a sail, seats, three regular oars, one steering oar, a **figurehead**, and a flag.

Mast

Sail

Seats

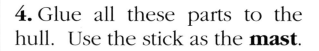

Steering oar

Oars

4. Glue all these parts to the hull. Use the stick as the **mast**.

5. Now cut out two squares of cardboard to make a stand for your longboat. Fold each piece in half and cut a hole in the middle for the boat to sit in.

Figurehead

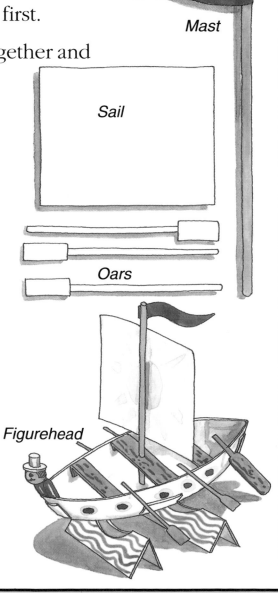

6. Decorate the boat with paint.

Making balsa wood boats

You will need:

A sheet of balsa wood   Scissors
A junior hacksaw   Wooden sticks
Thin cardboard   A straw

1. Cut out some small boats from the wood. Each boat should have a different **bow**.

2. Cut out some triangles and squares from the cardboard. These are the sails.

3. Use the sticks as masts. Thread the sticks through the sails. Make small holes in the boats to stand the masts in. Attach the sails to the boats.

4. Float your boats in water. Blow through the straw on the back of the sails. Which boat shape works best? Which sail shape works best?

5. Attach your best sail to your best boat. Do these two together work best of all?

> ### WARNING:
> When cutting balsa wood, always ask an adult to help you. Try not to breathe in any of the sawdust. Ask an adult to help you if you need to use a knife.

Making keels and rudders

You will need:

Flexible plastic
Scissors

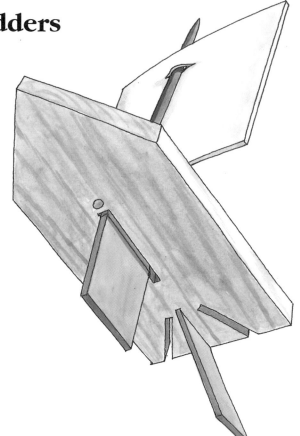

1. Use flexible plastic from a food container. Cut out a **keel** and a **rudder**.

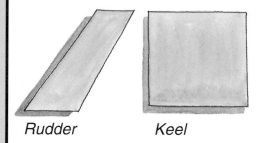

Rudder   *Keel*

2. Cut a slot in the middle of your balsa wood boat, big enough for your keel. The keel will help keep the boat upright and stop it from blowing sideways.

3. Cut three slots in the **stern** of the boat. Slide your rudder into one of the slots.

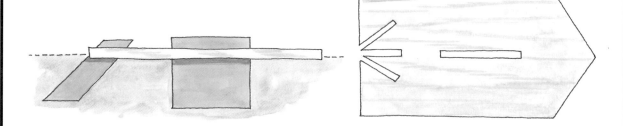

4. Float your boat in water. Put the rudder in different slots. In what direction does it move?

PADDLEBOATS

Paddle wheels can be used to make boats move along in the water. Have you ever ridden in a paddleboat? A paddleboat has a huge wheel with paddles that go around and around, propelling the boat forward.

Old paddleboats like this one take tourists on river trips.

Making a paddle boat

You will need:

Balsa wood
A junior hacksaw
A strong rubber band
A stapler

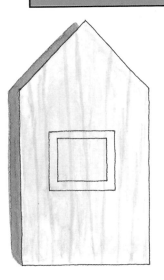

1. Cut out the shape of a boat from the balsa wood.

2. Cut a square piece out of the center of the boat. This is your paddle.

3. Make the square piece smaller than the hole so that it will fit into the hole with room to spare.

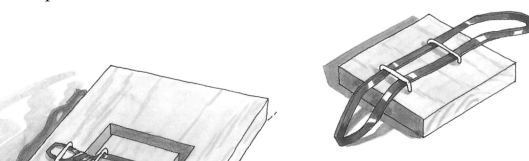

4. Staple the center of the rubber band to the paddle. Now staple the ends of the rubber band to the boat.

5. Wind up the rubber band by turning the paddle. Lower the boat into some water and let the paddle go. What happens?

MOTORBOATS

Making a motorboat

You will need:

Balsa wood
A junior hacksaw
A balloon

Plastic tubing
Tape
A strong rubber band

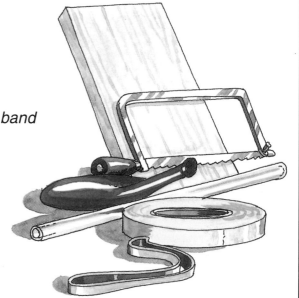

1. Cut out a balsa wood boat. Ask an adult to help you use the junior hacksaw.

2. Now stretch the mouth of the balloon over the end of the plastic tubing. Tape the balloon to the tubing.

3. Slide the end of the tubing under a rubber band at the back of the boat. Now blow up your balloon.

4. Put the boat in water and let the air out of the balloon. Watch your motorboat glide across the water.

Further work

Make plastic keels and rudders for your paddleboat and motorboat, like the ones you made on page 15. Attach a rudder to the front of one of your boats. Does a front rudder work as well as the rudder at the back?

Speedboats can go very fast. They make jets of water that push them along.

There are two main kinds of sail boats. Long ago, many big boats had sails that went across the boat. These were called square-riggers. Smaller boats, like racing yachts, usually have a fore-and-aft sail, or a sail that goes from the front to the back of the boat instead of from side to side.

*Look at all these racing yachts. Many of them have big extra sails at the front called **spinnakers** to make them go faster.*

Making a sailboat

You will need:

Two plastic bottles
Wooden sticks
Rubber bands
Paper
Scissors
Thread
Tape
Glue

1. Attach the two plastic bottles together using the wooden sticks and rubber bands.

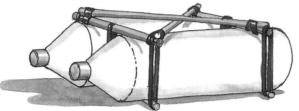

2. Cut out a triangle from paper to make a fore-and-aft sail.

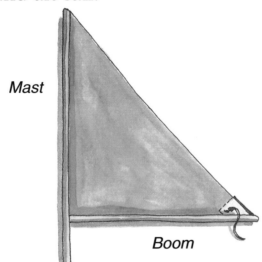

Mast

Boom

3. Use glue or rubber bands to make a mast and **boom** out of the sticks. Tape the sail to them.

4. Attach a short piece of thread to the corner of the sail.

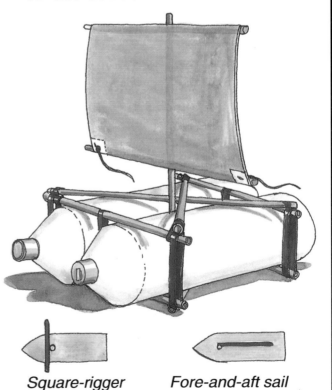

5. Place the lower end of the stick from your mast between the two bottles. Attach the free end of the thread to the end of the boat. Test your boat in water.

6. Now make a square-rigger sail in the same way. Attach it to your boat, as shown to the right. Which of the two sails works better?

Square-rigger   Fore-and-aft sail

PROPELLERS

A **propeller** has blades that spin around and push a boat through the water. You can make a propeller unit instead of a sail to use on your plastic bottle boat (see page 21).

Propellers like this one are used on big ships, such as oil tankers.

Making a propeller unit

You will need:

A stiff plastic tube
A bead
A nail
A cork
A paper clip
A strong rubber band
A model plastic propeller

1. Cut a piece of stiff plastic tube, about 9 1/2 inches (25 cm) long. You can get one from a craft store or hobby shop.

2. Straighten out the paper clip, leaving a hook at one end.

3. Thread the paper clip through the cork, the bead, and the propeller. Bend the straight end of the clip over the propeller.

4. Attach the rubber band to the hook at the other end of the paper clip. Pass the rubber band through the plastic tube. Use the nail to hold the rubber band at the other end, as shown.

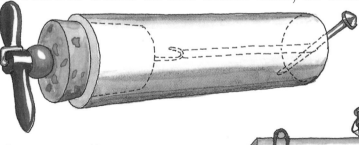

5. Attach the propeller unit to the back of your boat and wind up the propeller. Put the boat in water and let the propeller go!

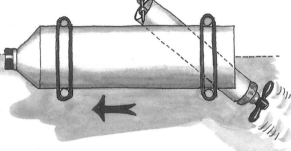

23

SUBMARINES

Do you ever wonder what is at the bottom of the sea? People often go deep into the water to build bridges or lay pipes. If they need to go very deep, they use submarines.

This huge submarine can stay under the water for many weeks at a time.

Making a model submarine

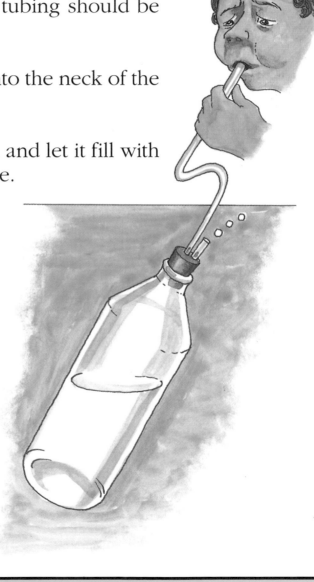

You will need:

A glass bottle
A rubber stopper with two holes in it
Plastic tubing
A large bowl of water

1. Push plastic tubes through the two holes in the stopper. One piece of tubing should be long and the other short.

2. Push the stopper firmly into the neck of the glass bottle.

3. Put the bottle in the bowl and let it fill with water through the short tube.

4. Now blow air through the long tube. Some of the water will be pushed out of the other hole in the stopper, and the bottle should rise.

5. Now put your finger over the end of the long tube. The air will stay in your "submarine." What happens when you take your finger off the end of the tube?

FISH

Fish live in water. Most fish have **fins** that help them balance and move in the water. If you have an **aquarium** at school or at home, look to see how the fish use their fins. Do they move their fins all the time?

There are many fish in this aquarium. Look at the shape and size of their fins.

Making a food chain mobile

Many different kinds of plants, fish, and tiny animals live in water. Some fish eat the tiny plants and animals in the water. Some big fish catch and eat small fish. A food chain shows how this works.

You will need:

Cardboard
Scissors
Felt-tip pens
String

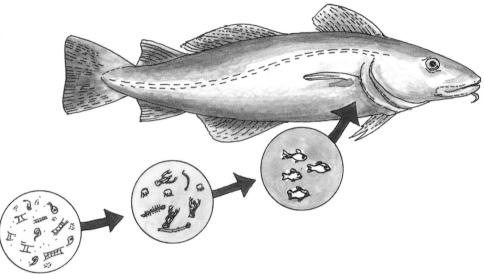

1. On cardboard, draw some animals and plants that live in water. You might draw a big fish, some small fish, tiny water animals, and masses of smaller water plants.

2. Cut out your pictures and string them together. Put them in the order of the food chain.

3. The fish at the top of the mobile eats many small plants and animals. So the large card in your food chain must be at the bottom, as shown.

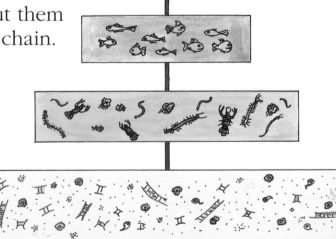

WHAT YOU'LL NEED

More Books About Water

Boat Book. Gail Gibbons (Holiday House)
Mississippi Sternwheelers. Pam Zeck and Gerry Zeck (Carolrhoda)
The Raft. Bess C. Haskell (Kennebec River Press)
Sailboat Racing. Claire Jones (Lerner)
Sailing. Norman Barrett (Franklin Watts)
The Ship Book. Michael Berenstain (David McKay)
Submarines. D. White (Rourke)
Think about Floating and Sinking. Henry Pluckrose (Franklin Watts)
Water, Water! Tom Johnston (Gareth Stevens)

More Books With Projects

Finding Out about Things that Float. A. Thomas (EDC Publishing)
Junior Science Book of Water Experiments. Rocco V. Feravolo (Garrard)
Micromodels: Make Your Own Six Little Ships. Myles K. Mandell (Putnam)
Model Boats and Ships. D. J. Herda (Franklin Watts)
Rub-a Dub-Dub, Science in the Tub. Jim Lewis (Meadowbrook)
Science Fun with Toy Boats and Planes. Rose Wyler (Julian Messner)
Simple Science Experiments with Water. Eiji Orii and Masako Orii (Gareth Stevens)
Young Scientist: The World of Water. Jerry DeBruin (Good Apple)

Places to Write for Science Supply Catalogs

Delta Education
P. O. Box M
Nashua, New Hampshire 03061

Ward's Natural Science
P. O. Box 1712
Rochester, New York 14603

Sargent-Welch Scientific Company
7300 North Linden Avenue
Skokie, Illinois 60076

Nasco Science
901 Janesville Road
Fort Atkinson, Wisconsin 53538

Schoolmasters Science
P. O. Box 1941
Ann Arbor, Michigan 48106

Macmillan Science Company
8200 South Hoye Avenue
Chicago, Illinois 60620

GLOSSARY

aquarium
A large glass container where live fish are kept.

boom
A long pole used to stretch out the bottom of a sail.

bow
The front part of a ship or boat.

figurehead
A carved figure attached to the front of sailing ships, used as a decoration.

fins
The thin parts standing out from the body of a fish to help it swim.

hull
The sides and bottom of a boat.

kayak
A small, narrow canoe holding only one person. It is closed except for an opening in the middle, where the paddler sits.

keel
A long piece of wood or metal attached to the bottom of a boat.

mast
A tall pole that holds up the sail on a boat.

outrigger
A float attached to the side of a boat to help keep the boat from tipping over.

propeller
A device with several flat blades sticking out from it, used to drive an aircraft or boat.

rudder
A flat piece of wood or metal attached to the back of a boat, used for steering.

spinnaker
An extra sail on the front of sailboats, used for moving very fast in the water.

spring balance
A device, like the hanging scales for produce in supermarkets, for measuring how much something weighs.

stern
The rear part of a ship or boat.

Vikings
People who lived in northern Europe about 1,000 years ago, famous for their long sea voyages.

Picture acknowledgements

The publishers would like to thank the following for allowing their photographs to be reproduced in this book: Eye Ubiquitous, p. 12; Tony Stone Worldwide, p. 10 (Chris Cole), 19, 20 (Alastair Black); Zefa, pp. 4, 8, 16 (H. Steenmans), 22, 24, 26 (C. Voight). Cover photography by Zul Mukhida.

INDEX